ELEMENTARY MATH
BEHIND
Google Maps

Why do we limit children's arithmetic skills to calculating candies or fruits!

- Nara Reddy

I am not affiliated with Google or Its products. I considered Google Maps to educate children.

ISBN: 979-8-8431-3908-7

Preface

Dear Parent / Teacher,

Thank you for considering this book.

Of course, there is no single approach that works for everyone, i would like to explain the approach I considered when I originally prepared these questions for my son.

My intention is not to up-skill him, as I believe teachers are the best at it. I wanted to help him relate his basic arithmetic skills to the life around him, and Google Maps seemed like a great fit since most of us use it so often in our lives. The questions will begin with simple one-step answers (like 5 x 2 = 10), and as they progress, they will require a sequence of addition and multiplication steps (sometimes subtraction) to find the answer.

Even though the questions often ask the student to calculate, I recommended to my son that he first figure out the step required and then he has the option to use a calculator or virtual assistant (Google Home, etc.) to perform the actual addition or multiplication. (Example: Once he understands it is 13 x 3, he can then use a calculator to find what 13 x 3 is).

The answers sections in this book provide detailed steps in case the student needs additional help to understand the process rather than just providing answers.

I hope your students/children will enjoy this book.

-   Nara Reddy
For questions or feedback: narabreddy@gmail.com

Acknowledgements

I would like to thank

A farmer and the first person who taught me basic elementary arithmetic, Babi Reddy, my father.

A math teacher who spent time to help me fix the language errors in this book, Alyssa Charles, Elmwood Franklin School.

Hi _____,

Have you ever seen Google Maps on your parent's phone? It guides us to drive to where we want to go.

See the picture, I am at Disney World and I want to drive to Jurassic World. Google Maps is showing two routes from Disney to Jurassic World.

Jurassic World is 16 miles away on the Blue Route and it takes 21 minutes to drive.

Jurassic World is 14 miles away on the Gray Route and it takes 27 minutes to drive.

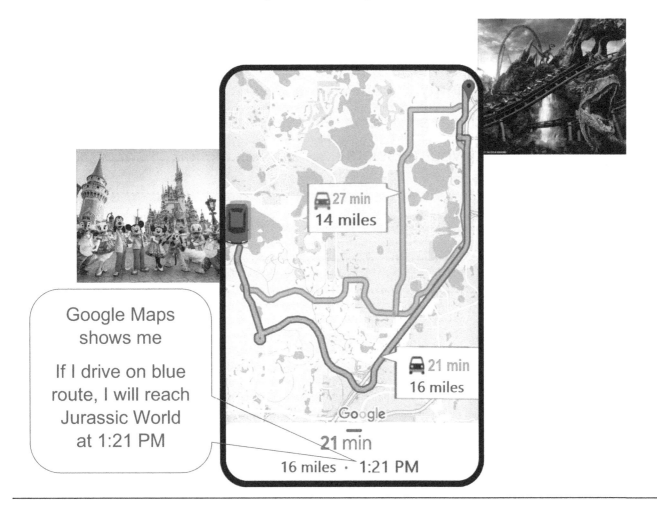

Google Maps shows me

If I drive on blue route, I will reach Jurassic World at 1:21 PM

Do you want to learn how Google Maps does this?

Table of Contents

Section 1: Are you ready? Let's go

1. I drove from my home to McDonald's to eat a sandwich. McDonald's is 1 mile away from my home.

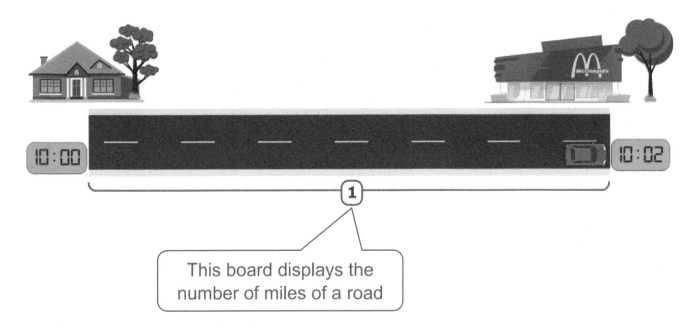

This board displays the number of miles of a road

It took 2 minutes to drive one mile to McDonald's. As you can see from the clock, I started at 10:00 o'clock and arrived at 10:02.

Remember, driving every mile takes 2 minutes

There is no need to answer anything on this page.

2. I would like to eat ice cream now, and the ice cream shop is 5 miles away from McDonald's.

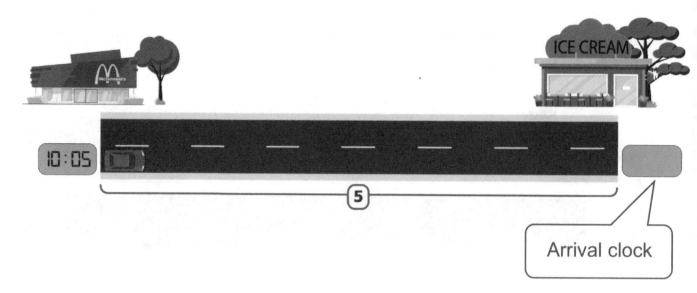

Arrival clock

a. How many minutes will it take to drive to the ice cream shop?

   Show how you calculate the answer in the orange box below.

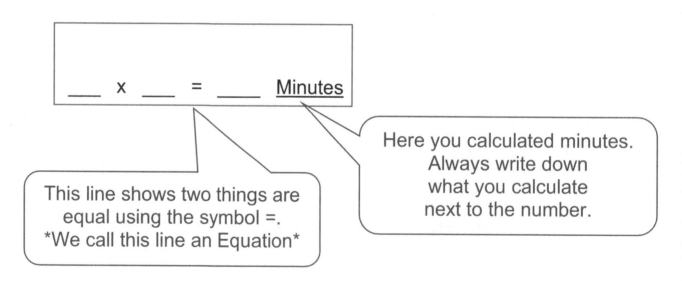

___ x ___ = ___ Minutes

This line shows two things are equal using the symbol =.
We call this line an Equation

Here you calculated minutes. Always write down what you calculate next to the number.

b. At what time will I reach the ice cream shop? Find and write it in the arrival clock.

3. After eating the ice cream, I decided to go to the park and play. The park is 13 miles away.

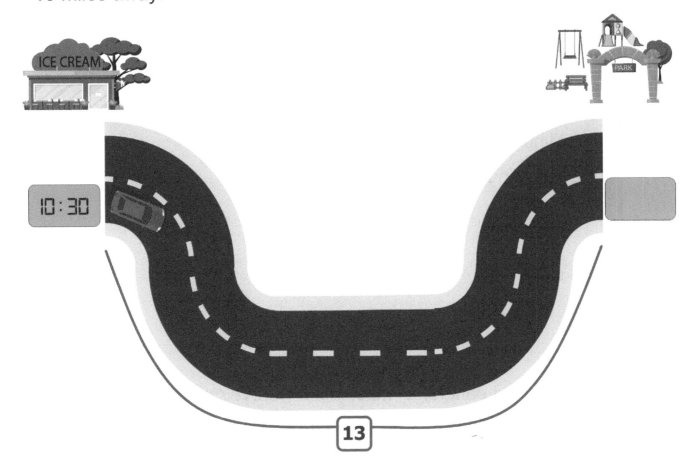

a. How long will it take to drive to the park? Calculate the number of minutes.

Remember to always calculate the answer with an equation.

b. At what time will I reach the park?

4. I drove on the local road until I reached the park, but from there I will have to drive on the highway. I can drive at high speed on the highway, which is faster than the local road.

How long does it take to drive 1 mile on the highway? Circle the correct answers from the options below in blue.

a. 1 Minute

b. Less than 2 Minutes

c. 2 Minutes

After playing in the park for a while, I finally drove on the highway and found out that, one mile took 1 minute.

Remember, driving every mile on the highway takes 1 minute

There is no need to answer anything here.

5. I am driving to the beach from the park. The beach is 18 miles from the park on the highway.

    a. How long will it take to drive to the beach? What time will I reach there?

6. My friend decided to go to the other beach from the park. You can see the highway from the park to the other beach.

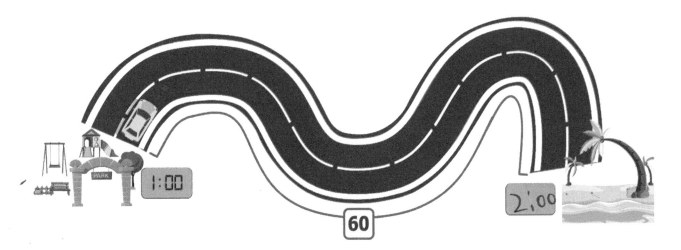

    a. How long will it take my friend to drive to the beach? What time will he reach the beach?

Have you observed how this book shows local roads and highways!

In each pointing bubble below, can you write down what type of road it is?

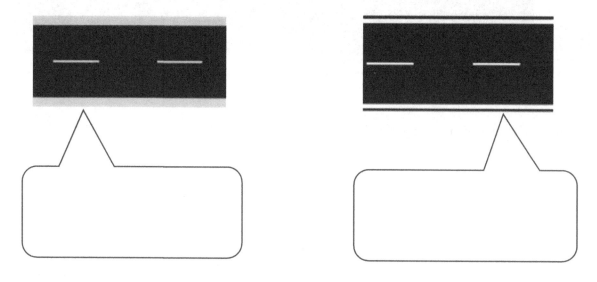

In this book, yellow lines will mark local roads and white lines will mark highways.

Remember to use the correct drive time when you calculate the time from now on.

Driving every mile on the highway takes 1 minute.
Driving every mile on the local road takes 2 minutes.

7. Vihan, a second-grade boy, missed his school bus one day. He will have to ride in his dad's car to school. See the road map below from Vihan's home to his school via 4 roads.

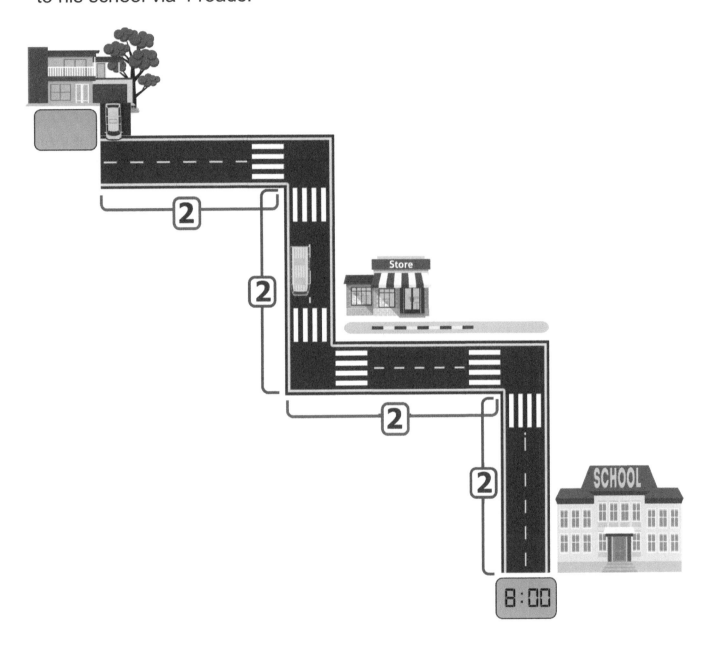

a. How long will it take for Vihan to reach school from his home? Calculate the time it takes to drive each of the four roads, and then calculate the total time.

> Calculate means
>
> adding or multiplying numbers in one or more steps to find the answer.
>
> Hint: All four roads have the same miles, so you will get the same equation 4 times before adding them in 5th equation.

b. You might have noticed the clock at school in the picture! Vihan must be at school by 8 A.M. What time should Vihan leave from home?

Please write the time in the departure clock at Vihan's home.

c. School ends at 3 PM. When does Vihan reach home if his dad picks him up from school at 3:15 PM?

8. Brian is good at soccer, so we call him a Soccer Boy. He plays soccer every Sunday, and his mom drives him to the soccer field. See the map below showing the route from his home to the soccer field via 4 roads.

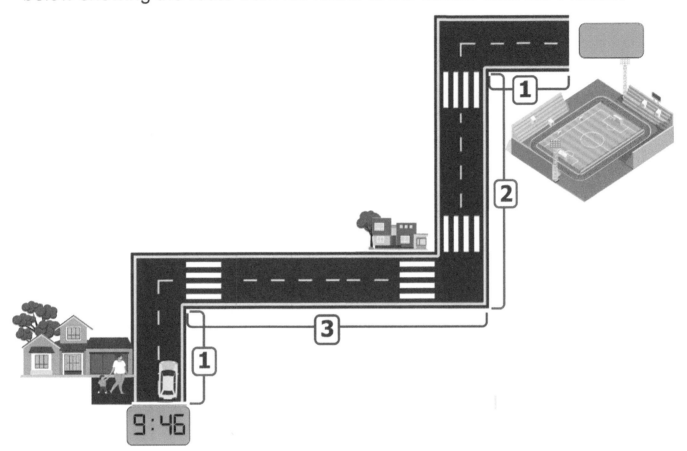

a. What is the distance between his home and soccer field? Calculate the total number of miles.

____ + ____ + ____ + ____ = ____ Miles

This equation calculates Miles.

b. How long will it take for his mom to drive from home to his soccer field? Use the distance you found above to calculate the answer.

c. Calculate the time it takes to drive each of the four roads, and then add them to find the total time.

d. Optional Question: Can you explain why the answer to the previous two questions (b and c) is the same?

9. Lucy loves tennis. Her dad drives her to tennis every Saturday. The following is a map showing the route from her home to the tennis court. You will notice two routes are available to reach the tennis court.

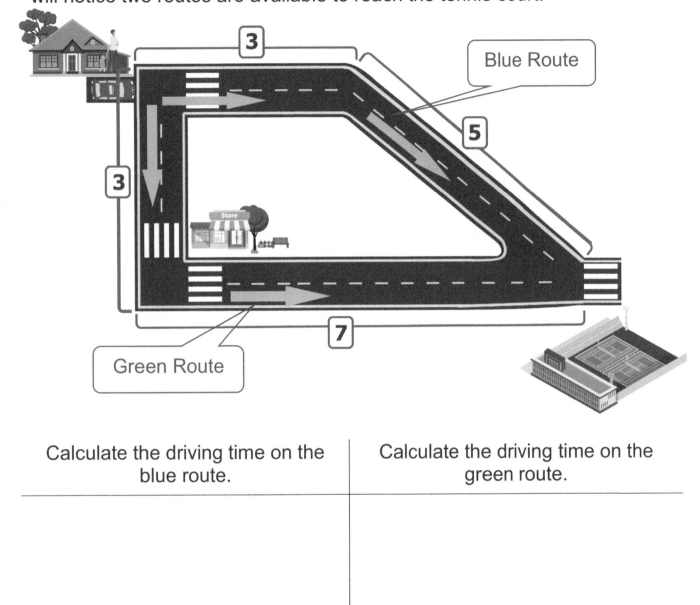

Calculate the driving time on the blue route.	Calculate the driving time on the green route.

Which route might Lucy's dad prefer to drive to get to the tennis court?

Hint: People usually choose the route that takes least amount of time.

10. Leo is an excellent hockey player, and he never likes to be late for practice. The map below shows Leo's dad's route to drop him off at his hockey rink.

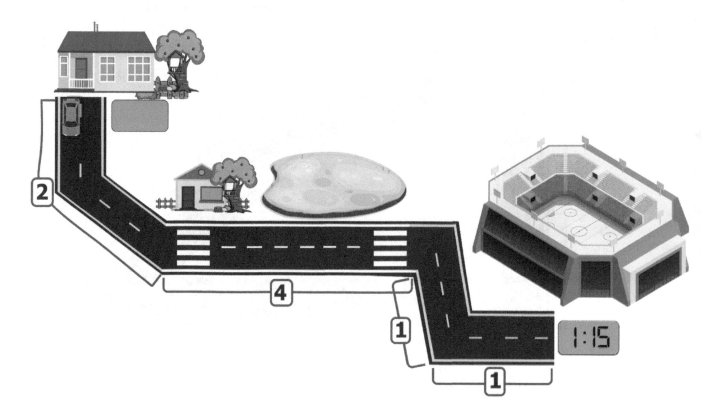

a. What is the distance between Leo's home and his hockey rink?

Highway Distance:

Local Road Distance:

Total Distance from home to the rink:

b. How long does it take to drive from his home to the rink?

Hint: Driving a mile on a highway and a local road takes different time

c. At the rink, the clock shows the next practice time. His practice starts at 1:15 PM. Leo is eating lunch at his house at 1:00 PM. Is he going to be able to make it to hockey practice on time?

d. If Leo will be able to reach hockey on time, please draw a smiley face.

e. If you found that he will be able to reach his hockey practise on time, explain under what condition.
   Tip: When is the latest he can leave to be on time for practice?

11. Emily enjoys swimming. The following is the map showing how to get from her home to her swim school.

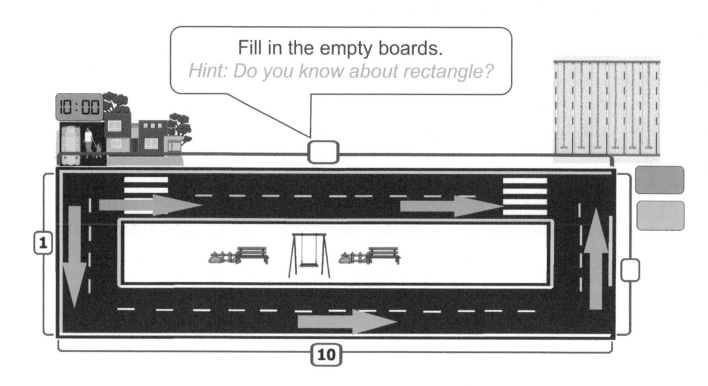

a. Emily's mom is ready to drive her to the pool. Calculate the distance from her home to the swimming pool for each route

b. Calculate the time it takes to drive each route.

c. What time do they arrive on the blue route and the green route? What route might Emily's mom prefer to take?

12. Iraj is a lean and athletic boy who does gymnastics well. The picture here shows the routes from his home to the gymnastics center.

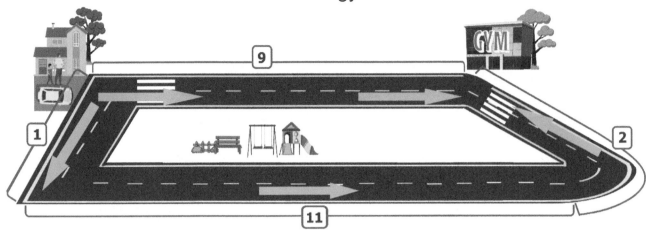

a. Which is the shortest route from his home to the gymnastics center? How many miles shorter is it?

Hint: The shortest route is the one with the least distance in miles.
You need to calculate the difference in miles to find how many miles shorter.

b. Which route should his dad take to get him to gymnastics faster?

Never guess the answer in math, always calculate and show the equations

c. How many minutes sooner does he reach the center compared to the other route?

13. Declan is a cool and smart boy who enjoys exploring different things in the world. He enjoys visiting museums and learning how things work.

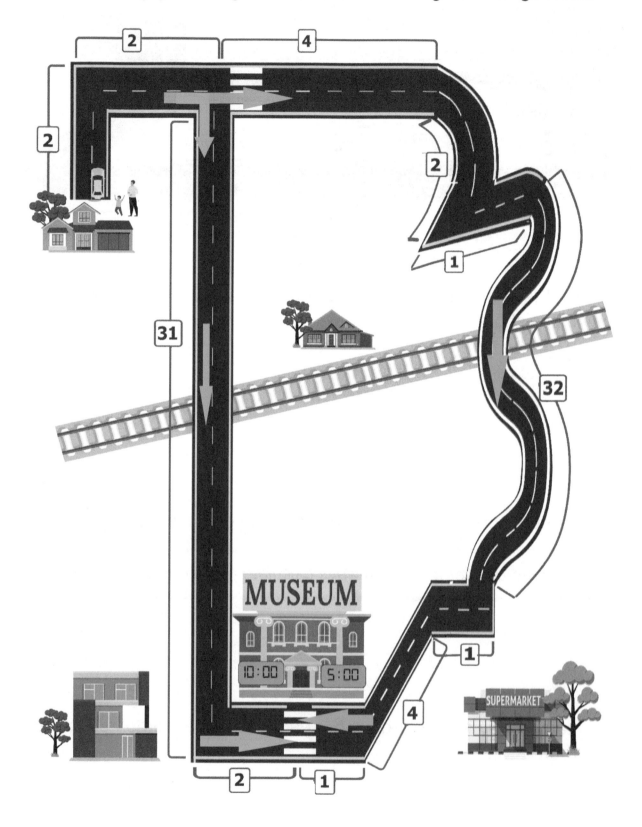

a. Declan's dad intends to drive route that is shortest distance. Which route is shorter? How many miles shorter?

b. He and his dad are leaving home at 9 o'clock. What time will they get to the museum on the shorter route?

c. Check out the museum's hours in the map, it's open from 10:00am to 5:00pm. If Declan wants to spend as much time as possible in the museum, could you please calculate the amount of time the other route takes to find the fastest route?

d. How much more time does he get to spend in the museum if his dad takes the faster route compared to the shorter route?

14. You have calculated car travel time up to this point.

    a. Can you name two vehicles that travel by air?

    b. Can you think of something that travels, but you cannot see? Write it below.

My teacher told me that sound travels. As you can see in the picture on the previous page, my voice travels from my mouth to other peoples' ears so they can hear me when I speak.

c. Have you ever wondered how fast sound travels?

Sound can travel 761 miles in an hour. It is about the same distance between New York City and Chicago.

d. Can you please ask your parent or teacher how long it takes to drive from New York to Chicago and write the answer below?

Hmmm…Sound will travel that distance in one hour.

There is a funny boy who can run 761 miles in an hour as well.

Guess Who

Guess Who

15. If Sonic runs 761 miles in an hour, will he be able to reach the moon in 13 days? *Hint: The first step is to find the number of hours in 13 days.*

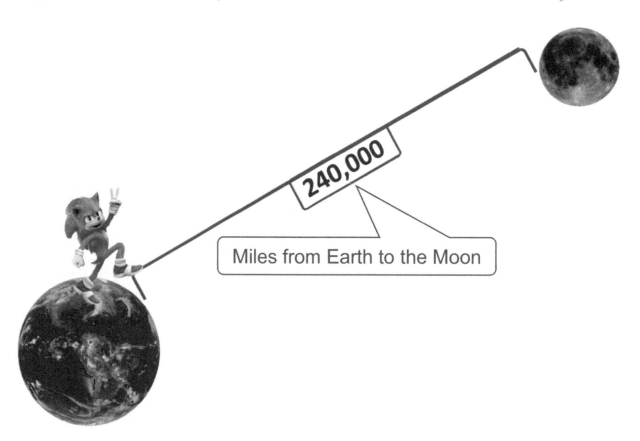

240,000

Miles from Earth to the Moon

Tip: You can use a calculator to work on big numbers

You Completed One Section.

Well Done

Section 2: The ride is going to be slippery

A. In the above picture, white powder is falling on the road. Do you know what it is?

B. Does it snow everywhere in the world in the winter?

C. Could you please write the names of two cities where it snows in the winter? *Hint: You can google or ask your parents or teachers*

D. Could you please write the names of two cities where it does not snow in the winter?

E. To reach our destination safely, should we drive faster or slower on the snowy road?

F. As we have seen, driving every mile on the local road took 2 minutes without snow. How long does it take to drive when the same road is covered with snow? Circle the correct answers.

        a. Less than 2 minutes

        b. 2 minutes

        c. More than 2 minutes

        d. 3 minutes

G. Driving every mile on the highway took 1 minute without snow. How long does it take to drive when the same road is covered with snow? Circle the correct answers.

        a. Less than 1 minute

        b. 1 minute

        c. More than 1 minute

        d. 2 minutes

1. On a snowy day, I again drove from my home to McDonald's to eat a sandwich.

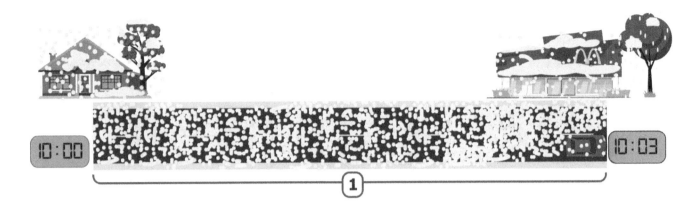

It took 3 minutes to drive one mile to McDonald's when the road was covered in snow. As you can see from the clock, I started at 10:00 o'clock and arrived at 10:03.

Remember, when there is snow on the road, driving every mile takes 3 minutes

---

2. After eating a sandwich, I would like to go to the ice cream shop and drink hot chocolate.

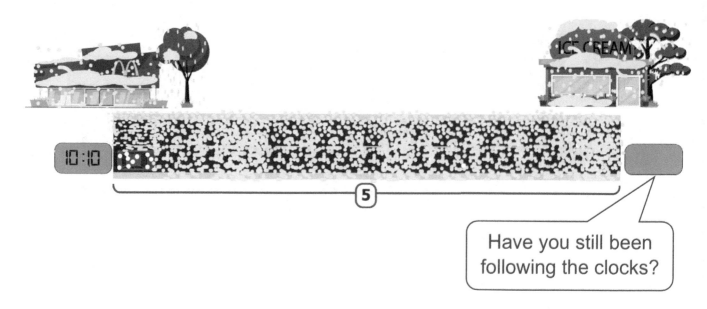

a. How many minutes will it take to drive to the ice cream shop?

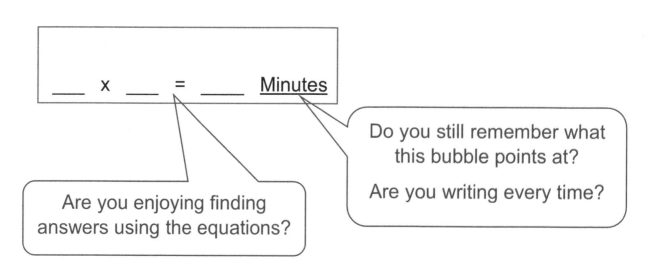

____ x ____ = _____ <u>Minutes</u>

b. At what time will I reach the ice cream shop?

3. After drinking hot chocolate, I decided to go to the park and see if there was snow there.

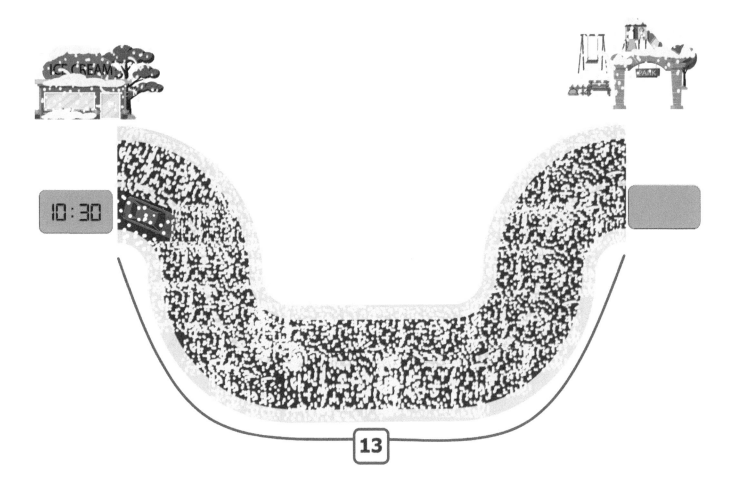

e. How long will it take to drive to the park?

f. At what time will I reach the park?

4. Seeing as there is so much snow everywhere, I have decided to go back home now. I have to drive home the same way I came to the park. How long does it take to reach my home?

There are two ways to find the answer to this question. Could you please show me how to calculate the time in two different ways?

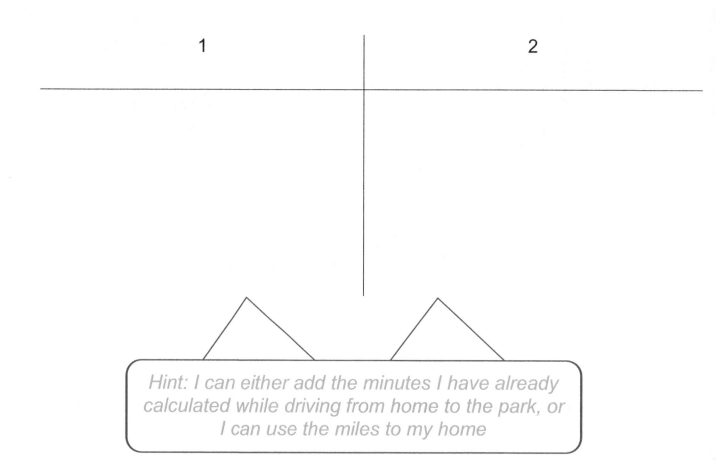

1

2

Hint: I can either add the minutes I have already calculated while driving from home to the park, or I can use the miles to my home

5. As you saw last time, my friend goes to another beach from the park. He thinks there is no snow on that beach, so he decided to drive there from the park.

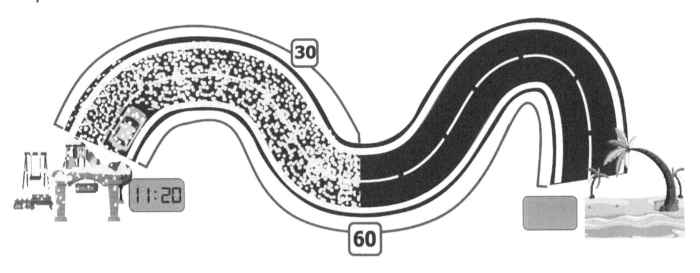

My friend told me that
Driving every mile on the highway takes 2 minutes
when there is snow on the road

a. Have you noticed that there is no snow on the road after 30 miles? How long will it take my friend to drive to the beach? What time will he reach there?

Fill in these blanks.

We found out that…

Without Snow
Driving every mile on the highway takes ___ minute.
Driving every mile on the local road takes ___ minutes.

With Snow
Driving every mile on the highway takes ___ minutes.
Driving every mile on the local road takes ___ minutes.

6. Vihan missed his school bus on the snowy morning. He will have to ride in his dad's car to school.

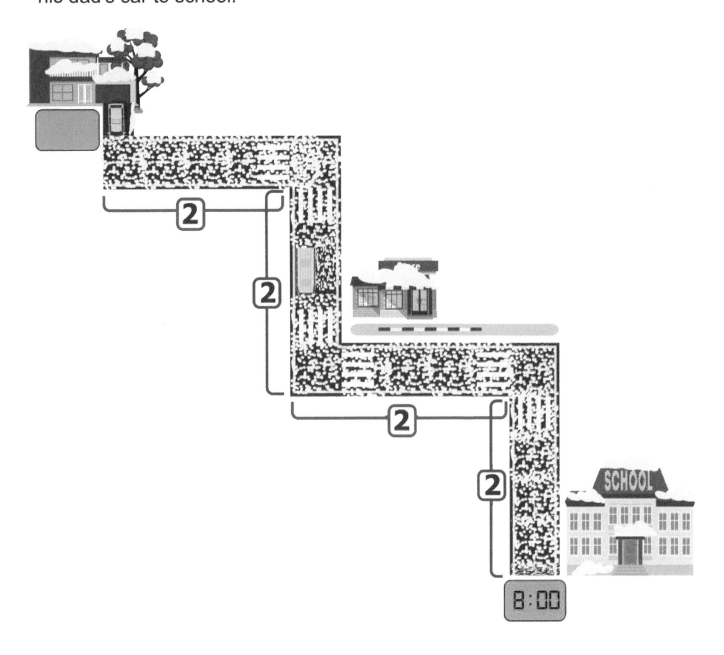

a. How long will it take for Vihan to reach school from his home? Calculate the time it takes to drive each road, and then the total time.

b. We know that his school begins at 8 A.M. What time should Vihan leave from home?

   Please write the time in the departure clock at Vihan's home.

c. School ends at 3 PM. When will Vihan reach home if his dad picks him up from school at 3 PM?

7. Brian enjoys playing soccer in the winter as well.

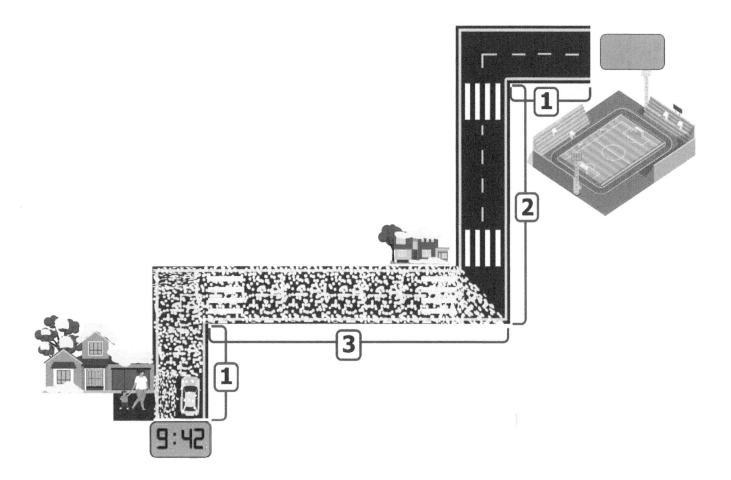

a. What is the distance between his home and the soccer field?

   Is it the same as what you have already calculated without snow, or should it be calculated again?

b. On the above route, how long will it take his mom to drive from home to his soccer field?

8. Lucy's tennis courts are indoors, so she likes to practice even on snow days. The snow has been plowed on one of the roads.

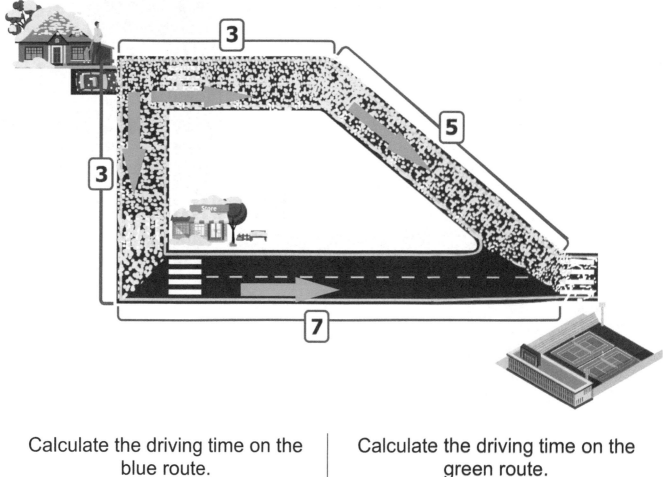

Calculate the driving time on the blue route.	Calculate the driving time on the green route.

Which route will Lucy's dad prefer to take to get to the tennis court?

Hint: People usually choose the route that takes least amount of time.

9. We know Leo never likes to be late for his hockey practice. Let's see if he can make it to the rink on time during the snowy day.

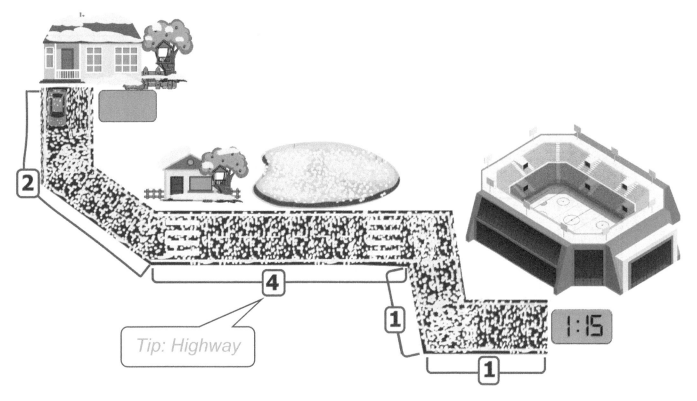

Tip: Highway

a. How long does it take to drive on the snowy road from his home to the rink?

b. His practice starts at 1:15 PM. Leo is eating lunch at his house at 1:00 PM. Is he going to be able to make it to hockey practice on time?

c. If you found that he will not be able to reach on time, how many minutes late will he be today if he leaves home at 1:05 PM?

10. Emily is ready to go to her swim class. This time, the route condition from her home is different.

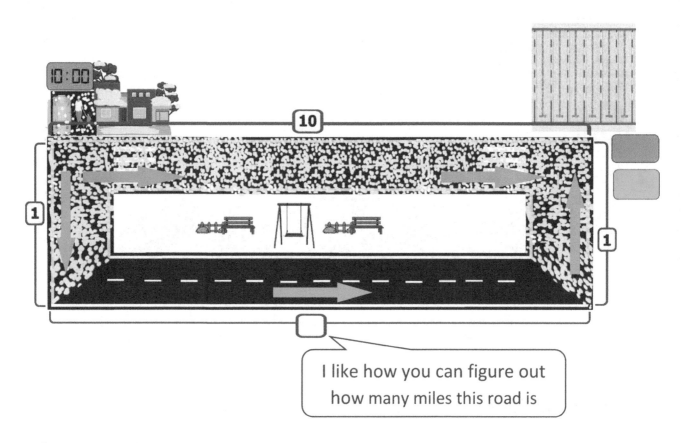

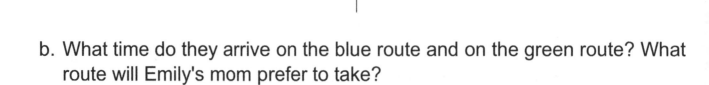

I like how you can figure out how many miles this road is

a. Can you calculate the time it takes to drive each route?

b. What time do they arrive on the blue route and on the green route? What route will Emily's mom prefer to take?

11. In the winter, Iraj wants to spend as much time as possible at the gymnastic center to stay fit for athletics. His goal is to go to the gymnastic center as soon as possible, but all the roads are covered in snow. Let's help his dad drive on the fastest route.

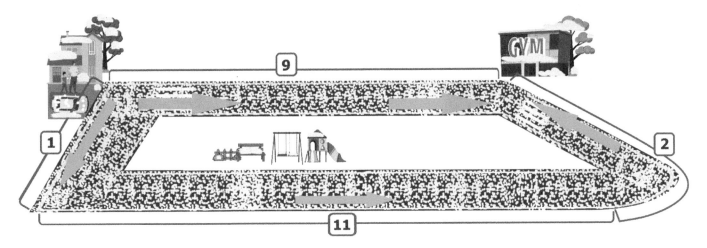

a. Which route should his dad take to get him to gymnastics faster?

b. How many minutes faster is the fastest route compared to the other route?

12. Let's see how much time Declan gets to spend in the museum this time.

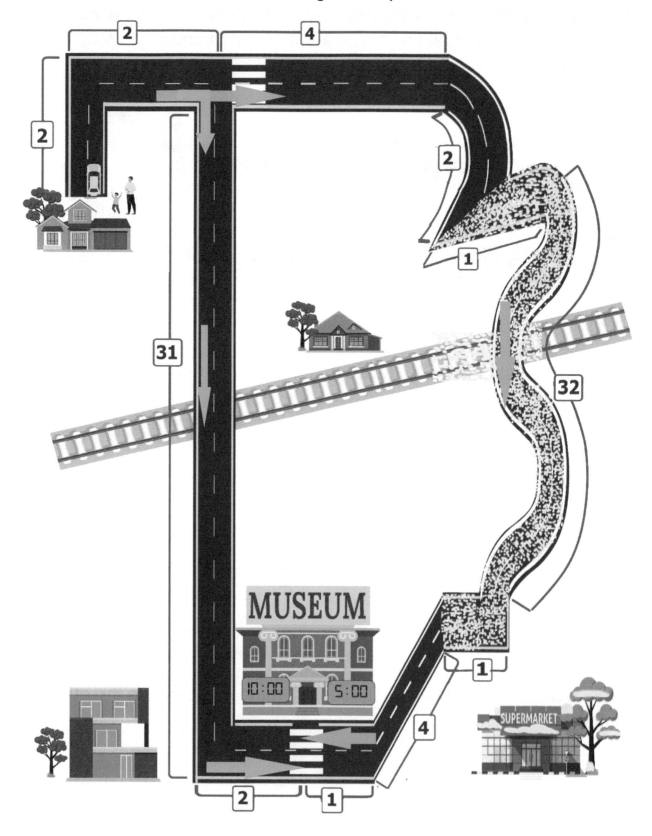

a. Last time, we calculated that the blue route was the faster route to the museum. As you see in the above picture, there is no snow near his home so his dad thinks he can take the blue route again to reach the museum faster. They are leaving home at **9 o'clock**. What time will they get to the museum on the blue route?

b. Find the time it takes if the green route is taken. Which route is the fastest?

c. How much more time does Declan get to spend in the museum if his dad takes the faster route?

13. In the previous section, we learned that some things are invisible, but they travel.

    a. Is there anything that travels, and you can feel its force even though you cannot see or hold it?

https://clipartion.com/free-clipart-19012/

Have you figured it out?

Have you ever felt the force of the wind?

My teacher told me that wind travels all the time, so I wondered why it isn't windy every day!

I was reading a website about wind on my iPad and found it very interesting. If you are interested in reading about wind, please ask your parent to open the website for you.

https://education.nationalgeographic.org/resource/wind

Scan the QR code to open the website on iPad or Tablet

b. What happens when the wind travels more than 74 miles per hour over the water?

Hint: Answer is available on the above website under the section **Results of Wind**.

You Completed Second Section.

High Five Please

Section 3: Enjoy the smiles on the road

https://commons.wikimedia.org/wiki/File:Traffic_congestion_at_A325.JPG

The heavy traffic on the road causes traffic congestion, which you can also call road congestion.

In traffic congestion, vehicles move slowly, delaying travel time.

We get time to see each other and smile

a. In the above picture, which side of the road do people drive on? Is it on their right-hand side or their left-hand side?

b. In the above picture, which side of the road do people drive on?

These pictures were taken in two different countries.

In the picture on the previous page, people drive on the right side of the road, while in the above picture, people drive on the left side.

Different countries have different road rules.

Below is a world map showing the right-hand side driving countries in red and the left-hand side driving countries in blue.

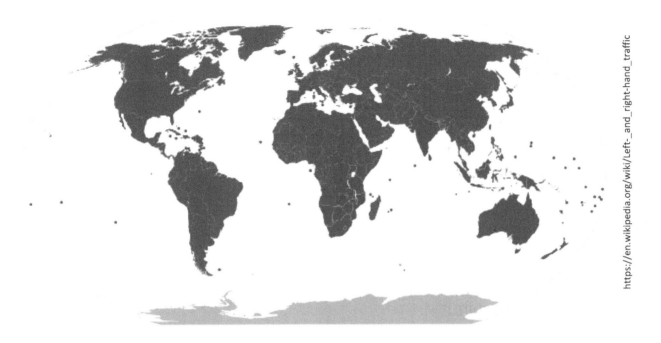

https://en.wikipedia.org/wiki/Left-_and_right-hand_traffic

c. Can you name three countries that drive on the right-hand side?

Tip: You can see country names on Google Maps and refer to their color in the above map.

d. Can you name three countries that drive on the left-hand side?

e. Driving every mile on the traffic-free local road took 2 minutes. How long does it take to drive in congested traffic?

    a. Less than 2 minutes

    b. 2 minutes

    c. More than 2 minutes

    d. 3 minutes

f. Driving every mile on the traffic-free highway took 1 minute. How long does it take to drive in congested traffic?

    a. Less than 1 minute

    b. 1 minute

    c. More than 1 minute

    d. 2 minutes

1. On a holiday, I drove to McDonald's to eat a sandwich and noticed that there were a lot of people driving on that road, causing congestion. Look at the traffic on the way to McDonald's.

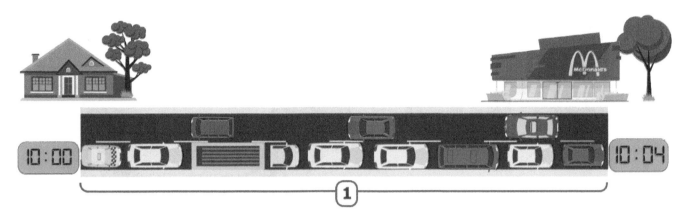

It took me 4 minutes to get there. I started at 10:00 o'clock and arrived at 10:04.

Remember, on a congested road, it takes 4 minutes to drive every mile

---

2. Due to traffic, I decided to skip the ice cream today and drive straight to the park.

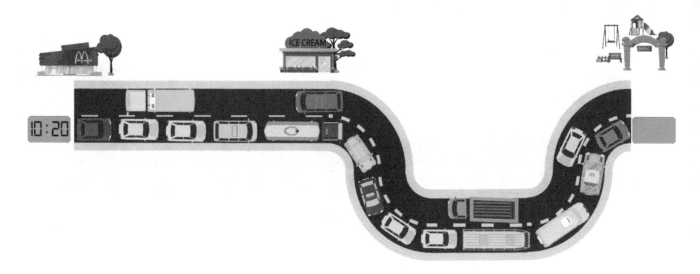

   a. What is the distance from McDonald's to the park?

   b. The entire route is congested as you can see in the map above. At what time will I reach the park?

3. From the park, I would like to go to the beach. On a sunny day, there are a lot of people who want to go to the beach, so the highway is also congested with traffic.

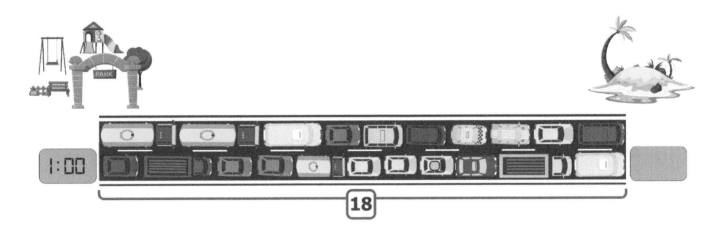

When traffic is congested, I have noticed that highways take longer than local roads. For every mile, it is taking 5 minutes to drive.

Remember, driving every mile on the congested highway takes 5 minutes

---

a. Today's weather forecast shows the sun setting at 6 PM. I am worried that I might not reach the beach before sunset. Will I arrive at the beach before sundown?

b. How much time do I get to spend at the beach before the sun sets?

4. I began driving back home when the sun set and there is no traffic on the roads. Could you please calculate the time at which I will arrive at my home?

There are two ways to find the answer to this question. Could you please show me how to calculate it in two different ways?

1	2
Have you calculated the traffic-free time for the same road previously?	

We know that…

With clear roads
Driving every mile on the highway takes 1 minute.
Driving every mile on the local road takes 2 minutes.

With Snow
Driving every mile on the highway takes 2 minutes.
Driving every mile on the local road takes 3 minutes.

With Congested Traffic
Driving every mile on the highway takes ___ minutes.
Driving every mile on the local road takes ___ minutes.

Fill in these blanks.

You do not need to memorize these.

You can always refer to this page when calculating the travel time for the routes.

5. Vihan missed his school bus on a busy morning, and the traffic was different this time.

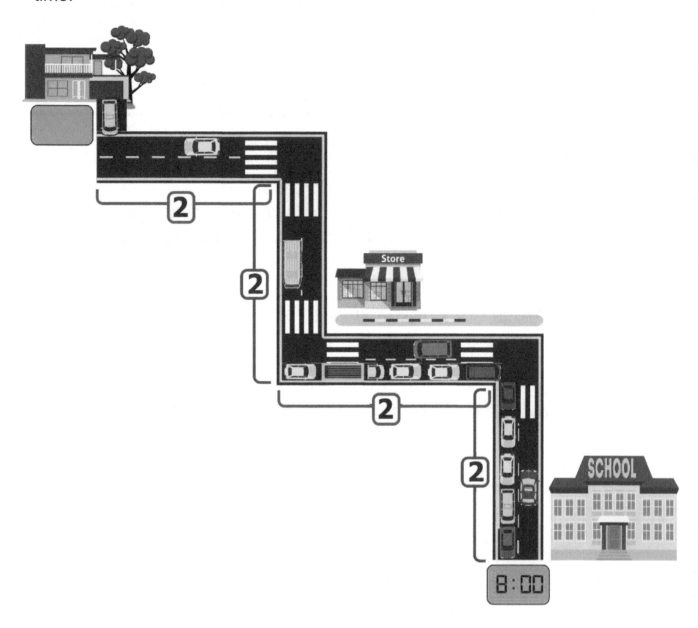

a. How long will it take for Vihan to reach school from his home? Calculate the time it takes to drive each road, and then the total time.

b. If Vihan leaves home at 7:36, will he reach school on time?

If Vihan reaches school on time, please draw a smiley face.

6. Here is a map showing the road conditions between Brian's home and his soccer field.

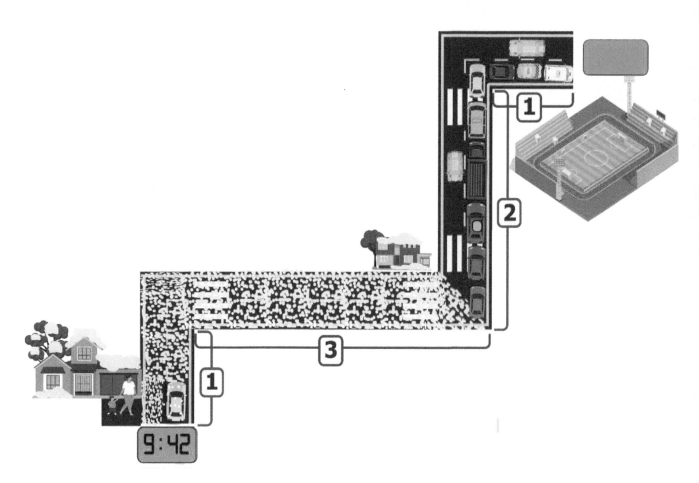

a. On the above route, how long will it take his mom to drive from home to his soccer field?

Are you still following, always calculate and show the equations?

---

7. Here is a map showing the road conditions between Lucy's home and her tennis court.

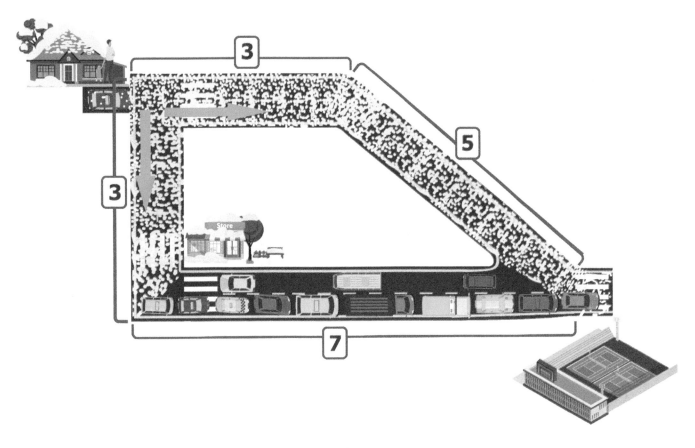

Calculate the driving time on the blue route.	Calculate the driving time on the green route.

Which route will Lucy's dad prefer to drive to the tennis court?

8. Leo's route has some unusual traffic conditions

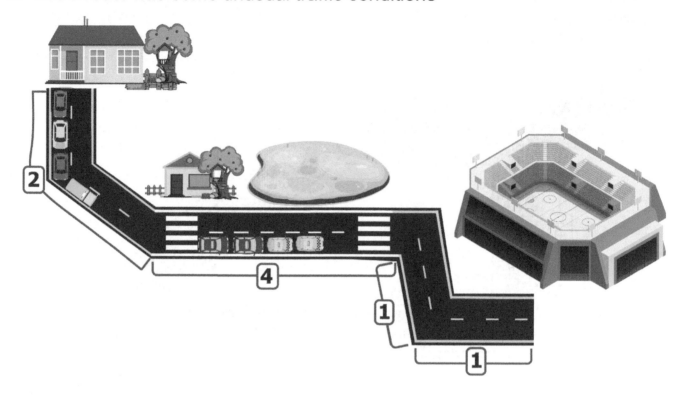

a. Can you calculate how long it will take his dad to drive him from home to the rink?

9. Emily is ready to go to her swim class, but if some roads are congested, her mom isn't sure if the green route is the fastest. We should find her the quickest route.

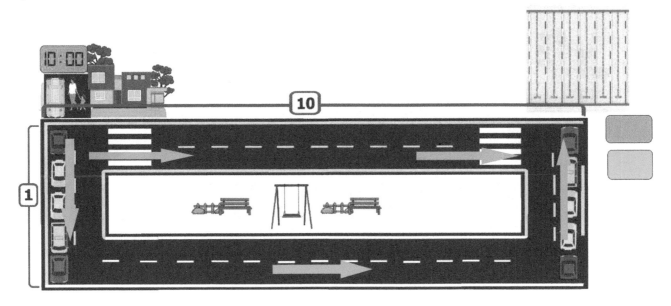

a. Can you calculate the time it takes to drive each route?

b. What time do they arrive on the blue route and on the green route? Which is the fastest route?

10. Iraj's dad uses the maps on his phone to find the fastest route, but it does not show him any route as the fastest this time, so he is confused about which route to take. I'm hoping you can help him find the fastest route.

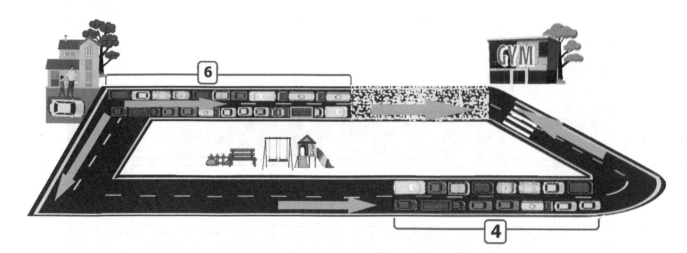

a. Can you calculate the time it takes to drive the blue route?

b. Can you calculate the time it takes to drive the green route?

c. What was the reason that the maps on his phone did not show any route as the fastest?

11. We know Declan's dad likes to drive on the shortest route, and Declan wants to explore the museum for as long as possible.

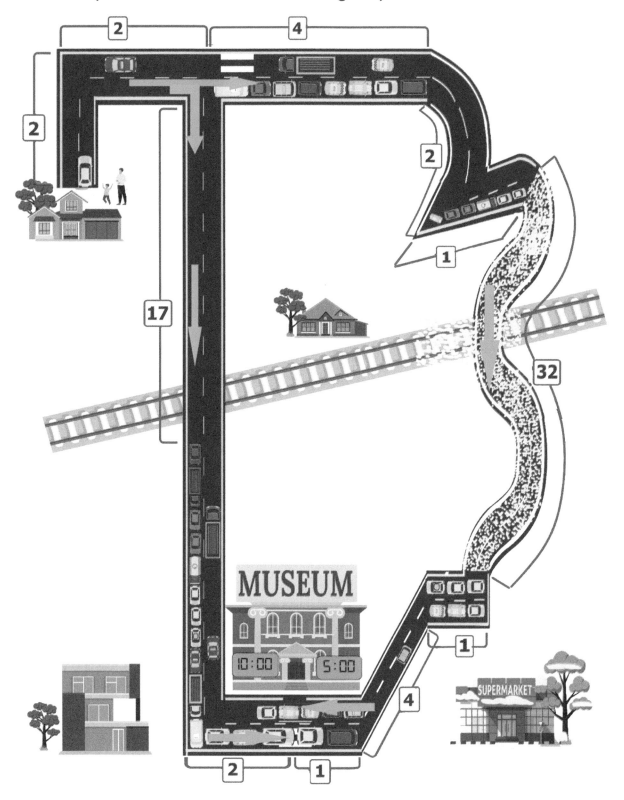

a. They are leaving home at 9 AM. What time will they get to the museum on the blue route?

b. What time will they get to the museum on the green route?

c. How much time does Declan get to spend in the museum today?

Hint: You can go on this website to calculate time duration once you know Declan's arrival time and the museum closing time.

https://www.calculator.net/time-duration-calculator.html

You Completed Third Section.

Great Job, Way to Go

Section 4:You're full of energy to continue

1. What does your body need to perform every physical activity?

image: clipartion.com

2. How does your body get its energy?

3. Could you please write two healthy foods you eat to get energy? We call them the source of your energy.

4. What is the source of a car's energy?

image: Flaticon.com

5. Different countries have different names for the fuel we use for our cars. Could you figure out what the colors in this map indicate?

What Gasoline is called around the World

■ Petrol ■ Gasoline ■ Benzene ■ Essence ■ Naphtha ■ Others

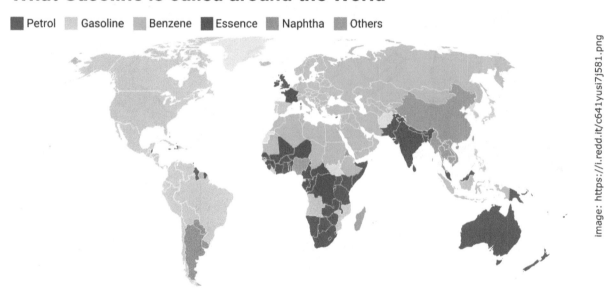

image: https://i.redd.it/c641yusi7j581.png

a. Can you name three countries where people call fuel, gasoline?

b. Can you name three countries where people call fuel, benzene?

6. Let us assume your parent needs one gallon of gas a day to drive to work and back home. The price of one gallon is $3.

> Gasoline is called gas

   a. If your parent drives to work 5 days a week, how many gallons of gas does he or she need each week?

   b. How much does your parent have to pay for gas each week?

   c. If your parent drives to work four weeks a month, how many gallons of gas does he or she need each month?

> *There are two ways to find the answer to this question. Could you please show how to calculate the answer in two different ways?*

1	2

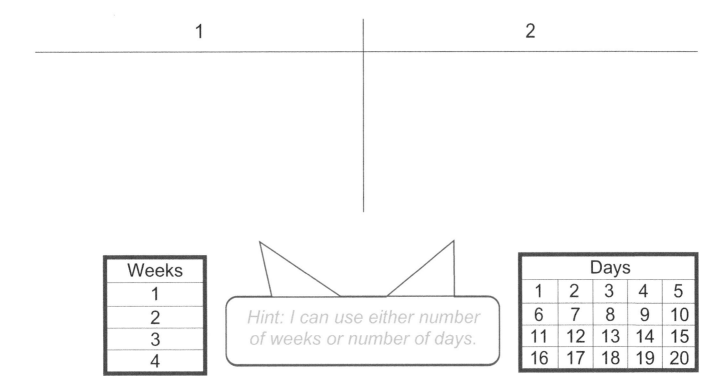

Weeks
1
2
3
4

Hint: I can use either number of weeks or number of days.

Days				
1	2	3	4	5
6	7	8	9	10
11	12	13	14	15
16	17	18	19	20

d. How much does your parent have to pay for gas each month?

If you would like, you can try two methods of calculating the answer!

1	2

e. How much gas does your parent need in a year if he or she drives to work 11 months out of the year?

f. How much does your parent have to pay for gas in a year?

g. My teacher told me that I could show all the answers I calculated above for gas and money in a single table like the one below. It would be great if you could understand and fill it out!

Gas	One Day	One Week	One Month	One Year
Number of Gallons	1			
Cost ($)	3			

7. You might know that there are electric vehicles that do not require fuel! What do you think they use for energy to run the motors in these cars?

image: clipartion.com

Have you heard that electricity is the greatest invention in human history?

Are you curious about how electricity was invented?

Here is a funny YouTube video to help you understand electricity.

https://www.youtube.com/watch?v=sPJjsE24So8

As you may have noticed, gasoline is measured in gallons, but electricity is measured in kilowatts.

Assume your parent drives an electric vehicle to work and it uses 10 kilowatts of electricity a day to drive to work and back home. It costs $1.

a. Could you please calculate all the values for this table and fill them in?

Electricity	One Day	One Week	One Month	One Year
Number of kilowatts	10			
Cost ($)	1			

You are the Champion

Be Proud of Yourself!

Let's have an interview please

What was it like working on this book? Did you have fun?

Would you like to explain what did you learn from this book?

What would you like to learn more about in your life?

I would be happy to read your answers. Would you like to email this page to me? Please ask your parent or teacher.

If you would like to verify your answers!

Section 1

2.

   a. *Correct Answer: 5 x 2 = 10 Minutes*

     2 x 5 = 10 Minutes gets you the same answer. But 5 x 2 is the right way to find the answer. Do you like to know why?

     The picture below shows a 5-mile road and that takes 2 minutes to drive each mile. Do you see 5 times 2 or 2 times 5?

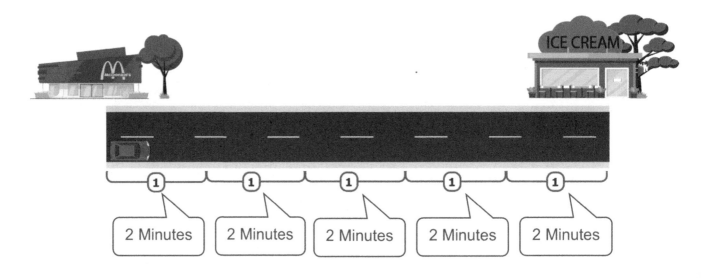

   b. *10:05 + 0:10 = 10:15*

3.

   a. *Correct Answer: 13 x 2 = 26 Minutes*

     2 x 13 = 26 Minutes gets you the same answer. But 13 x 2 is the right way to find the answer.

   b. *10:30 + 0:26 = 10:56*

Section 1

4. *Correct Answer: Less than 2 Minutes*

It is possible that answer 1 Minute is correct as well, but we won't know until we drive on the highway next.

5.

    a. *18 x 1 = 18 Minutes*

       1:00 + 0:18 = 1:18

6.

    a. *60 x 1 = 60 Minutes = 1 hour*

       1:00 + 0:60 = 2:00   or   1:00 + 1:00 = 2:00

7.

    a.

       2 x 2 = 4 Minutes
       2 x 2 = 4 Minutes
       2 x 2 = 4 Minutes
       2 x 2 = 4 Minutes

       4 + 4 + 4 + 4 = 16 Minutes   or   4 x 4 = 16 Minutes

    b. *8:00 – 0:16 = 7:44*

    c. *3:15 + 0:16 = 3:31*

8.

    a. *1 + 3 + 2 + 1 = 7 Miles*

Section 1

b. *7 x 2 = 14 Minutes*

c.

1 x 2 = 2 Minutes
3 x 2 = 6 Minutes
2 x 2 = 4 Minutes
1 x 2 = 2 Minutes

2 + 6 + 4 + 2 = 14 Minutes

9:46 + 16 = 10:00

Brian reaches soccer field at 10:00

9.

Blue route	Green route
3 + 5 = 8 Miles	*3 + 7 = 10 Miles*
8 x 2 = 16 Minutes	*10 x 2 = 20 Minutes*

Lucy's dad might prefer blue route.

10.
  a.

Highway Miles: 4
Local road Miles: 2 + 1 + 1 = 4

Total number of Miles: 4 + 4 = 8

  b.

Highway time: 4 x 1 = 4 Minutes
Local road time: 4 x 2 = 8 Minutes
Total time: 4 + 8 = 12 Minutes

c.

1:00 + 0:12 = 1:12

Yes, he can reach on time

d.

e. *1:15 – 0:12 = 1:03*

Condition: He must leave home before 1:03

11.

	Blue route	Green route
a.	*10 Miles*	*10 + 1 + 1 = 12 Miles*
b.	*10 x 2 = 20 Minutes*	*Local road: 2 x 2 = 4 Minutes* *Highway: 10 x 1 = 10 Minutes* *4 + 10 = 14 Minutes*
c.	*10:00 + **0:20** = 10:20*	*10:00 + **0:14** = 10:14*

Emily's mom might prefer green route.

12.

a.

Blue route: 9 Miles
Green route: 1 + 11 + 2 = 14 Miles

Blue route is the shortest route.

14 – 9 = 5 Miles

Blue route is 5 miles shorter than green route

b.

Blue route: 9 x 2 = 18 Minutes

Green route:
    3 x 2 = 6 Minutes
    11 x 1 = 11 Minutes

    6 + 11 = 17 Minutes

Green route is the fastest.

c.   *18 – 17 = 1 Minute*

13.

a.

Green route: 2 + 2 + 31 + 2 = 37 Miles

Blue route: 2 + 2 + 4 + 2 + 1 + 32 + 1 + 4 + 1 = 49 Miles

Green route is shorter.

49 – 37 = 12 Miles
Green route is 12 miles shorter

Section 1

b. *Green Route: 37 x 2 = 74 Minutes*

c. *Blue Route:*

   Local Road: 11 x 2 = 22 Minutes

   Highway: 38 x 1 = 38 Minutes

   Total time: 22 + 38 = 60 Minutes

Blue route takes lesser time than green route, so blue is the fastest.

d. *74 – 60 = 14 Minutes*

By taking the blue route, he arrives 14 minutes earlier, therefore, he has 14 more minutes to spend compared to the green route.

14.   *I hope you figured out the answers*

15.

1 Day = 24 Hours
13 Days = 13 x 24 = 312 Hours
312 x 761 = 237,432 Miles

237,432 is less than 240,000 miles, so he will not reach the moon in 13 Days.

Hey, did you try checking if Sonic can reach the moon in 14 days?

Section 1

Section 2

We drive slowly on the snow road to reach safely.

A. *Snow*

B. *No*

C.
D.
E.
   I hope you found out the answers for C, D, E

F. *Correct Answer: More than 2 Minutes*

   It is possible that answer 3 Minutes is correct as well, but we won't know until we drive on the snow road.

G. *Correct Answer: More than 1 Minute*

   It is possible that answer 2 Minutes is correct as well, but we won't know until we drive on the snow road.

2.
   a. *Correct Answer: 5 x 3 = 15 Minutes*

      3 x 5 = 15 Minutes is not better answer as we have seen in Section 1.

   b. *10:10 + 0:15 = 10:25*

3.
   a. *Correct Answer: 13 x 3 = 39 Minutes*

      3 x 13 = 39 Minutes is not better answer.

   b. *10:30 + 0:39 = 11:09*

4.

39 + 15 + 3 = 57 Minutes	*13 + 5 + 1 = 19 Miles*
	19 x 3 = 57 Minutes

5.

a.

*Snow Road: 30 x 2 = **60 Minutes***
Distance without Snow: 60 – 30 = 30 Miles
*30 x 1 = **30 Minutes***

*Total time to reach the beach: **60 + 30** = 90 Minutes = 1 Hour 30 Minutes*

11:20 + 1:30 = 12:50

6.

a.

2 x 3 = 6 Minutes
2 x 3 = 6 Minutes
2 x 3 = 6 Minutes
2 x 3 = 6 Minutes

6 + 6 + 6 + 6 = 24 Minutes   or   4 x 6 = 24 Minutes

b. *8:00 – 0:24 = 7:36*

c. *3:00 + 0:24 = 3:24*

Section 2

7.

   a. *Distance will not be affected by snow. It is 7 Miles*

   b.

      Snow Road: 4 x 3 = 12 Minutes
      Road without snow: 3 x 2 = 6 Minutes
      Total amount of time = 12 + 6 = 18 Minutes

      9:42 + 18 = 10:00

      Brian reaches soccer field at 10:00

8.

Blue route	Green route
3 + 5 = 8 Miles *8 x 3 = 24 Minutes*	*3 x 3 = 9 Minutes* *7 x 2 = 14 Minutes* *9 + 14 = 23 Minutes*

   Lucy's dad might prefer green route.

9.

   a.

      Highway time: 4 x 2 = 8 Minutes
      Local road time: 4 x 3 = 12 Minutes
      Total time: 8 + 12 = 20 Minutes

   b.

      1:00 + 0:20 = 1:20

      He will not be able to reach on time

Section 2

c.

1:05 + 0:20 = 1:25
1:25 – 1:15 = 0:10

He will be 10 minutes late.

10.

	Blue route	Green route
a.	*10 x 3 = 30 Minutes*	*2 x 3 = **6** Minutes* *10 x 1 = **10** Minutes* ***6** + **10** = 16 Minutes*
b.	*10:00 + 0:30 = 10:30*	*10:00 + 0:16 = 10:16*

Emily's mom might prefer green route.

11.

a.

Blue route	Green route
9 x 3 = 27 Minutes	*3 x 3 = **9** Minutes* *11 x 2 = **22** Minutes* ***9** + **22** = 31 Minutes*

Iraj's dad should take blue route to get him faster.

b.

31 – 27 = 4 Minutes

Section 2

12.

   a. *Blue Route:*

        *No Snow Local Road: 9 x 2 = **18** Minutes*
        *Snow Local Road: 2 x 3 = **6** Minutes*
        *No Snow Highway: 6 x 1 = **6** Minutes*
        *Snow Highway: 32 x 2 = **64** Minutes*

        *Total time: **18 + 6 + 6 + 64** = 94 Minutes = 1 Hour 34 Minutes*

     9:00 + 1:34 = 10:34

     They will get to museum at 10:34

   b. *No Change in the green route, answer is same as what you have calculated in previous section. Green route takes 74 Minutes*

   c. *94 – 74 = 20 Minutes*

13.

   a. *Wind*

   b. *Hurricane (Have you ever seen a hurricane in person or on TV?)*

Section 2

Section 3

a. *Right Side*

b. *Left Side*

c. *I hope you pointed out country names for c and d. Good job.*
d.

e. *Correct Answer: More than 2 Minutes*

   It is possible that answer 3 Minutes is correct as well, but we won't know until we drive on a congested road. Check out question 1 in this section.

f. *Correct Answer: More than 1 Minute*

   It is possible that answer 2 Minutes is correct as well, but we won't know until we drive on a congested road. Check out question 3 in this section.

2.

a. *5 + 13 = 18 Miles (I hope you remembered I drove on the same route in previous sections)*

b. *18 x 4 = 72 Minutes = 1 Hour 12 Minutes*

   10:20 + 1:12 = 11:32

3.

a. *18 x 5 = 90 Minutes = 1 Hour 30 Minutes*

   *1:00 + 1:30 = **2:30***

   *I will reach at **2:30**. I will reach before sunset.*

b. *6:00 - **2:30** = 3:30 = 3 Hours 30 Minutes*

4.

*18 + 26 + 10 + 2 = **56 Minutes***	*Highway: 18 x 1 = 18 Minutes* *Local Road: 19 x 2 = 38 Minutes* *18 + 38 = **56 Minutes***

*6:00 + **0:56** = 6:56*

I will reach home at 6:56

5.

   a.

     No-traffic Road: 2 x 2 = 4 Minutes
     No-traffic Road: 2 x 2 = 4 Minutes
     Congested Road: 2 x 4 = 8 Minutes
     Congested Road: 2 x 4 = 8 Minutes

     4 + 4 + 8 + 8 = 24 Minutes

   b. *7:36 + 0:24 = 8:00*

     He will reach on time.

6.

   a.

     Snow Road: 4 x 3 = 12 Minutes

     Traffic Road: 3 x 4 = 12 Minutes

     12 + 12 = 24 Minutes

Section 3

7.

Blue route	Green route
3 + 5 = 8 Miles *8 x 3 = 24 Minutes*	*3 x 3 = 9 Minutes* *7 x 4 = 28 Minutes* *9 + 28 = 37 Minutes*

Lucy's dad might prefer blue route.

8.

   a. *The miles are not given for the traffic section, we cannot calculate the time. Oh man, I may have tricked you* 😊

   If you tried any other way, very nice try.

9.

	Blue route	Green route
a.	*10 x 2 = 20 Minutes*	*2 x 4 = 8 Minutes* *10 x 1 = 10 Minutes* *8 + 10 = 18 Minutes*
b.	*10:00 + 0:20 = 10:20*	*10:00 + 0:18 = 10:18*

Green route is the fastest.

Section 3

10.

  a.

    Traffic road: 6 x 4 = 24 Minutes

    Snow road:

        9 – 6 = 3 Miles

        3 x 3 = 9 Minutes

    Blue route: 24 + 9 = 33 Minutes

  b.

    No-Traffic local road: 3 x 2 = 6 Minutes

    Traffic highway: 4 x 5 = 20 Minutes

    No-Traffic highway:

        11 – 4 = 7 Miles

        7 x 1 = 7 Minutes

    Green route: 6 + 20 + 7 = 33 Minutes

  c. *Both the routes take same time*

Section 3

11.

   a. *Blue Route:*

       No-Traffic Local Road:

$$2 + 2 = 4 \text{ Miles}$$
$$4 \times 2 = \textbf{8 Minutes}$$

       Traffic Local Road:

$$4 + 1 + 1 + 1 = 7 \text{ Miles}$$
$$7 \times 4 = \textbf{28 Minutes}$$

       No Traffic Highway:

$$4 + 2 = 6 \text{ Miles}$$
$$6 \times 1 = \textbf{6 Minutes}$$

       Snow Highway: $32 \times 2 = \textbf{64 Minutes}$

       Total time: $\textbf{8 + 28 + 6 + 64}$ = *106 Minutes = 1 Hour 46 Minutes*

   9:00 + 1:46 = 10:46

   They will get to museum at 10:46 AM

   b. *Green route*

       No-Traffic Local Road:

$$2 + 2 + 17 = 21 \text{ Miles}$$
$$21 \times 2 = \textbf{42 Minutes}$$

       Traffic Local Road:

$$31 - 17 + 2 = 16 \text{ Miles}$$
$$16 \ \times 4 = \textbf{64 Minutes}$$

       Total time: $\textbf{42 + 64}$ = *106 Minutes = 1 Hour 46 Minutes*

   Same as Blue route.

Section 3

c.

His dad can take any route, Declan will reach museum at 10:46 AM and museum will close at 5:00 PM.

5:00 PM – 10:46 AM = 6:14 = 6 Hours 14 Minutes

Have you tried calculating the time duration on this website?

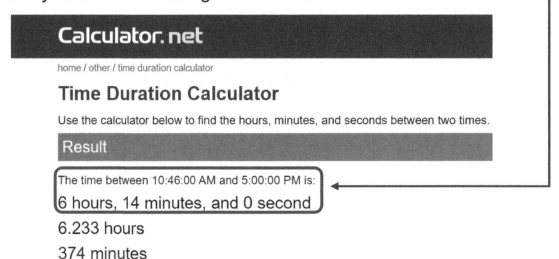

_____

It took me a long time to calculate the answers to question 11 and it was hard.

If you tried it, good job.

_____

Section 3

Section 4

1. *I think it is Energy. What do you think?*

2. *I get my energy from the healthy food I eat and drink. How do you get?*

3.

4. *Fuel.*

   If you have written any other answer for car's energy, you know better than I do.

5. *I hope you pointed out country names for c and d. Good job.*

6.
   a. *5 x 1 = 5 Gallons*

   b. *Correct answer: 5 x 3 = $15*

      3 x 5 = $15 is not better answer. Can you find why?

Monday	Tuesday	Wednesday	Thursday	Friday
$3	$3	$3	$3	$3

Section 4

c.

1	2
4 Weeks in a Month 5 Gallons each Week 4 x 5 = 20 Gallons	5 Days in a week 4 Weeks = 4 x 5 = 20 Days 1 Gallon each day 20 x 1 = 20 Gallons

d.

1	2
20 Gallons each month Each gallon costs $3 20 x 3 = $60	$15 each week 4 Work weeks each month 4 x 15 = $60

e.

20 Gallons each month
11 work months in a year
11 x 20 = 220 Gallons

f.

$60 each month
11 work months in a year

11 x 60 = $660

Section 4

g.

Gas	One Day	One Week	One Month	One Year
Number of Gallons	1	5	20	220
Cost ($)	3	15	60	660

7.

a.

Electricity	One Day	One Week	One Month	One Year
Number of kilowatts	10	50	200	2200
Cost ($)	1	5	20	220

Section 4

Made in the USA
Las Vegas, NV
09 July 2023